Introducing 'Claws' – the Surrey dinosaur

The finding of this enormous claw-bone in 1983 led to the discovery of the fossilized remains of a new kind of flesh-eating dinosaur. It turned out to be the most exciting and important dinosaur find in Britain or Europe this century. 'Claws', as it was soon nicknamed, attracted world-wide interest from scientists and in the Press and on radio and television.

The dinosaur was found in rocks about 124 million years old from the Wealden stage of the Early Cretaceous period. The remains of flesh-eating dinosaurs are comparatively rare, and are often very fragmentary. This new animal is the only reasonably complete skeleton of a large flesh-eating dinosaur found in Britain this century, and it is the only one from rocks of Early Cretaceous age anywhere in the world.

The claw-bone – it measures 31 centimetres along the outside curve.

How the skeleton was found

The story began one day in January 1983, when Mr William Walker, an amateur collector, obtained permission from the Ockley Brick Company to look for fossils in their claypit in Surrey. He noticed, lying in the mud, a rugby ball-shaped rock from which a small fragment of bone protruded. Mr Walker picked up the rock, shattered it with his geological hammer, and out fell several fragments of an enormous claw-bone. He took the pieces home, reassembled them and then discovered that the tip of the claw-bone was missing. Mr Walker went back to the same spot some days later and, incredibly, found the missing tip, still encased in rock, exactly where it had fallen from his hand.

Mr Walker's son-in-law, also an amateur fossil collector, brought the claw-bone to the Natural History Museum to show it to the scientists. They realized that it was an exciting and unusual find, and that it belonged to a flesh-eating dinosaur, but at that stage they had no idea how important it would turn out to be.

Mr William Walker with his find.

At the site

The story of the discovery of *Baryonyx* is one of great good fortune. First of all Mr Walker just happened to be in the right place at the right time. It was lucky that he found the claw-bone, rather than some less spectacular bone such as a rib, for that alerted the scientists to something very unusual. It was fortunate that bulldozers had scraped away the overlying clay and exposed the bones, and it was very fortunate indeed that the machines had not dug six inches deeper or the bones would have been completely destroyed. Mr Walker did the best possible thing and alerted the Museum quickly. Had he not done so, the clay-digging season would have restarted in the spring and the dinosaur would undoubtedly have been ground up and made into bricks!

Mr Walker accompanied two of the Museum's palaeontologists on an inspection of the site in February 1983. Together they discovered a lot more bone just beneath the surface of the clay, in spite of the wet and muddy winter conditions. The spot was close to where the claw-bone was found, right next to the main roadway in the claypit. Later in the spring of the same year (1983), when the weather had improved and the clay had dried out, the Museum sent a team of people to excavate the dinosaur remains. The task took eight people three weeks to complete.

Some of the Museum team in the early stages of the excavation in the claypit, spring 1983.

The 'crazy paving' of blocks containing the bones.

The plan of the 'crazy paving'. Although the skeleton was broken up most of the bones had not moved far from their natural positions. Skull bones (circle); vertebral column and ribs (square); shoulder girdle and front legs (triangle); hip girdle and hind legs (diamonds).

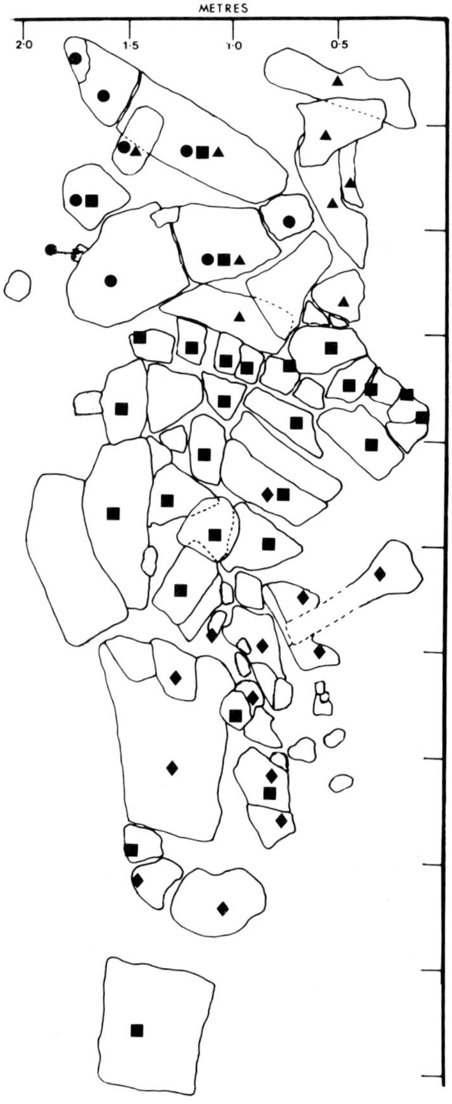

Excavating the bones

Most of the dinosaur bones were preserved in nodules of hard siltstone, surrounded by Weald Clay (used to make bricks). Every nodule contained one or more pieces of bone. Some of the bones, however, were found lying loose in the clay. The clay had first to be removed with hand tools such as trowels, putty knives and wood chisels, exposing the separate blocks of rock. Picks and shovels were used to excavate a wide trench around this 'crazy paving' in order to check for further fossil bone. Wherever bone was exposed it was cleaned carefully using metal modelling tools, knives and paint brushes. Then the bone received a coat or two of consolidant (varnish) to strengthen it, and any breaks were repaired. Eventually the whole surface was cleaned up ready for photographs and sketches to be made. The position, size and orientation of the blocks and bones were carefully noted on a plan; each block was also numbered and the numbers were recorded on the plan.

Each block was separated from its neighbours by excavating the clay from cracks in the 'crazy paving' and from beneath. In order to protect the fossil bone where it was exposed, the blocks were cocooned in plaster of Paris bandages (after covering the bone with tissue paper to act as a separator) or expanded polyurethane foam (with aluminium cooking foil as the separator). Thus protected, 54 blocks weighing over two tonnes in total were transported safely to the Natural History Museum.

Applying plaster of Paris bandage to an excavated block in order to protect the dinosaur bones during transport to the Museum.

Excavating around a block containing bones protected by a partial cocoon of expanded polyurethane foam.

In the laboratory – 1

Opening the cocooned fossil blocks was not difficult although it could be a messy process. Initially only half of the cocoon (generally the original underside) was cut away, leaving the fossil supported in the remaining half. Normally this was done using hand tools such as saws, knives, wood chisels etc., but after the installation of a shot blaster (a 'Vacu-Blast Nova 150PB') the cocoon and the underlying clay were blasted away in a tenth of the time taken by the manual method. Any fossil bone exposed by these methods was further cleaned with a dental machine (a 'Prophyjet') which jets water and sodium carbonate powder at high pressure. After rinsing and drying, the fossil bone was strengthened by applying a resin solution.

Removal of the hard rock was, and still is, the major problem. The siltstone is heavy and tough, and made more so by the presence of the mineral siderite, an iron carbonate. Although dilute thioglycollic acid will dissolve iron salts, the process is very slow and the siltstone still has to be removed by mechanical preparation. Hammer and chisels are still used today to remove rock quickly, especially modern tungsten carbide-tipped chisels, but more up-to-date equipment can produce better results in a shorter time.

The right upper arm bone being prepared from its cocoon of expanded polyurethane. The stone masons' chisels and lump hammer can be seen at the top of the picture, with two wood chisels used to remove softer clay and plastic foam.

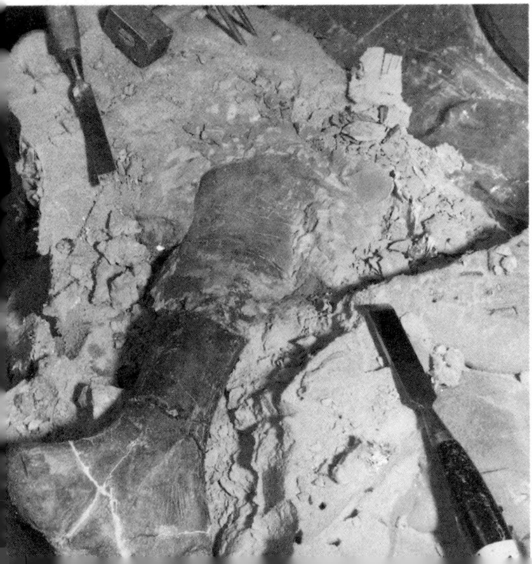

Preparing the fragile bones of the lower jaw with hammer and chisel.

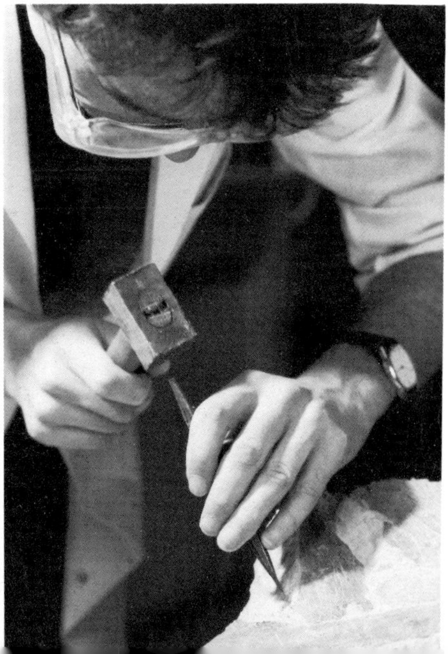

In the laboratory – 2

The bulk of the rock is now, in general, efficiently removed with a shot blaster. Powered by compressed air and operating inside a protective cabinet, it fires a stream of abrasive particles which eats away the rock at a great rate. As bone appears it is protected by rubber latex which, on drying, repels the abrasive. Large chisels and diamond-coated circular saws are also driven by compressed air. The saws are used to cut grooves in the rock, and it is then easy to chisel away the ridges left behind. Working closer to the fossil bone requires more care, and a 'micro-sandblaster' and finely pointed air-powered engravers are used, generally under a stereo-microscope.

After it has been prepared, fragile bone is 'hardened' or consolidated with a synthetic resin solution and repairs are made with fast-setting synthetic resin cements. The bones are stored in the Department of Palaeontology's specially designed and air-conditioned storage to protect them permanently.

Enough of the Surrey dinosaur has now been prepared to allow the publication of a preliminary scientific description of the animal. However, several more years of skilled and time-consuming preparation are required before the dinosaur can be fully described and exhibited. Moulding and casting of the bones is already under way to provide other museums with replicas, and, possibly, to mount the whole skeleton for future exhibition.

Preparing the snout and its teeth with an air-powered engraver.

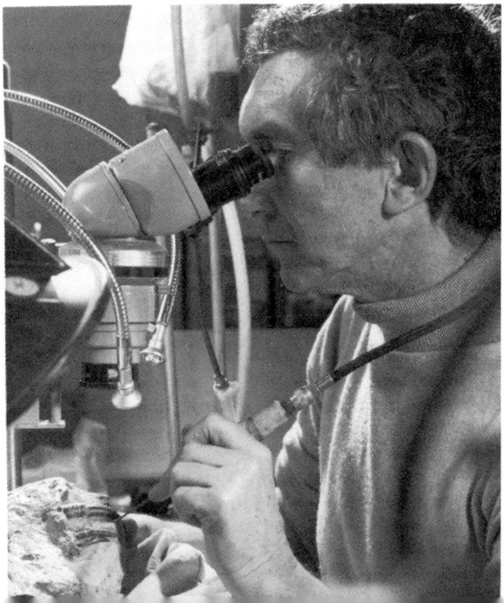

A silicone rubber mould and a glass-fibre and resin cast of the claw-bone.

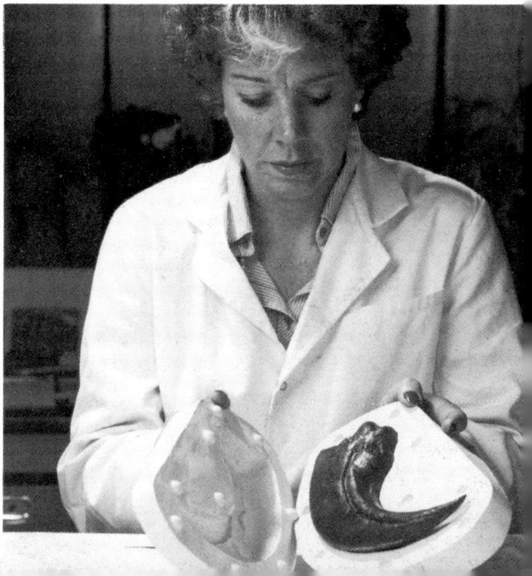

Naming the beast

All animals and plants, living or fossil, have a two-part scientific name based on Latin or Greek. This rule is followed in every country so that scientific names are recognized and understood throughout the world without a language barrier. The scientific name may described some unique feature of the organism, or indicate where it was found or who found it.

'Claws' scientific name is *Baryonyx walkeri*. *Baryonyx* is from the Greek meaning 'heavy claw' which describes the enormous claw-bone, and *walkeri* is in honour of Mr William Walker, who found the claw which led to the discovery of the skeleton.

Although the skeleton was discovered in 1983, it was only after several of the bones had been prepared that the scientists were able to determine that they belonged to a new kind of dinosaur. 'Claws' was officially named in November 1986 and attracted renewed interest around the world – it even starred in its own television programme in February 1987.

NATURE VOL. 324 27 NOVEMBER 1986

Baryonyx, a remarkable new theropod dinosaur

Alan J. Charig & Angela C. Milner

Department of Palaeontology, British Museum (Natural History), Cromwell Road, London SW7 5BD, UK

An extremely large claw bone, some 30 cm long, was found in Wealden (Lower Cretaceous) deposits in a Surrey claypit in January 1983. This led to the discovery the following month of the well-preserved skeleton of a new large theropod dinosaur. Only one other theropod specimen comprising more than a few bones had ever been found in Britain, and that discovery was more than a century ago. Indeed, no large theropod, reasonably complete, had previously been discovered in Lower Cretaceous rocks anywhere in the world. Our study so far suggests that the Surrey dinosaur was a typical large theropod in certain respects, resembling, for example _Allosaurus_[1]. In several other respects, however, it differs sufficiently from all known dinosaurs to merit designation as the representative of a new species, genus and family.

Family **Baryonychidae** nov.

Diagnosis. As for genus *Baryonyx*, below.

Genus *Baryonyx* nov.

Derivation of name. Gk, *βᾰρύ*, heavy, strong; *ὄνυξ*, talon, claw.

Type-species. *B. walkeri* nov.

Diagnosis. As for species *B. walkeri*, below.

Species *Baryonyx walkeri* nov.

Derivation of name. For Mr William J. Walker, who found the first indication of the skeleton, namely the large claw bone (ungual phalanx).

Holotype. B.M.(N.H.) Palaeo. Dept. R. 9951.

'Claws' named. (Courtesy of *Nature*).

How is *Baryonyx* related to other dinosaurs?

Baryonyx is a flesh-eating dinosaur. It belongs to the same group, the Theropoda, as all other flesh-eaters (such as *Deinonychus, Allosaurus, Megalosaurus* and *Tyrannosaurus)*, although it is sufficiently different from all of those to merit its own family – the Baryonychidae.

Baryonyx hits the headlines...

What did *Baryonyx* look like?

Bones from all parts of the animal were embedded in the blocks of rock. Enough have been prepared so far to make up about half of the skeleton. It can be pieced together by comparing the bones with those of other dinosaurs – rather like doing a giant three-dimensional jigsaw puzzle. Many of the bones that are missing or damaged can be restored by matching them against their opposites, for example missing parts of the right hind leg can be drawn or modelled as mirror images of the bones from the left leg. Unfortunately, most of the tail was missing but the few bones that were found indicate that *Baryonyx'* tail would have been similar to those of other large flesh-eaters. Muscles and skin are restored around the framework of bones to give an idea of the appearance of the animal in life. It was about 9 or 10 metres long, 3 – 4 metres tall in bipedal pose and probably weighed around 2 tonnes.

a

b

c

Baryonyx had a narrow skull with a long, low snout, a long neck and large powerful arms. The large claw-bone, which is 31 cm long, was probably borne on the inside finger (the thumb) of the hand. In life the claw-bone would have been covered with a horny sheath – just like a claw of a bird – so the whole claw would have been perhaps 35 cm long. *Baryonyx* probably walked on its hind legs like other theropods. Its front legs were relatively longer and more robust than those of other theropods so it might have been able to walk on all fours too – which no other theropod could do.

The bones of *Baryonyx* found so far (a) including the jaws (b), from which the skeleton (c) can be reconstructed, and then a restoration be produced of its possible appearance on two legs (d) or four (e).

d

e

The life style of *Baryonyx* – fishing or scavenging?

Baryonyx almost certainly ate fish, at least as part of its diet. Several features of the skull suggest this. The snout is elongated with a spoon-shaped tip, and the upper jaw is S-shaped in side view. The lower jaw is very slender and lightly built. It contained 32 teeth on each side, twice the number found in most other flesh-eating dinosaurs. Modern fish-eating crocodiles have jaws shaped rather like those of *Baryonyx,* with a curving line and a large number of teeth to snatch and grip their slippery prey.

Teeth and scales of a Wealden fish called *Lepidotes* have been found inside the rib cage of *Baryonyx.* The scales appear to have been partly digested in the stomach, so they may represent *Baryonyx'* last meal. *Baryonyx* might have fished rather like grizzly bears today. It may have crouched on river banks or waded into shallow water and hooked fish with its large claws, or snatched

The head of the *Baryonyx* (a) compared with that of a fish-eating crocodile (b).

a

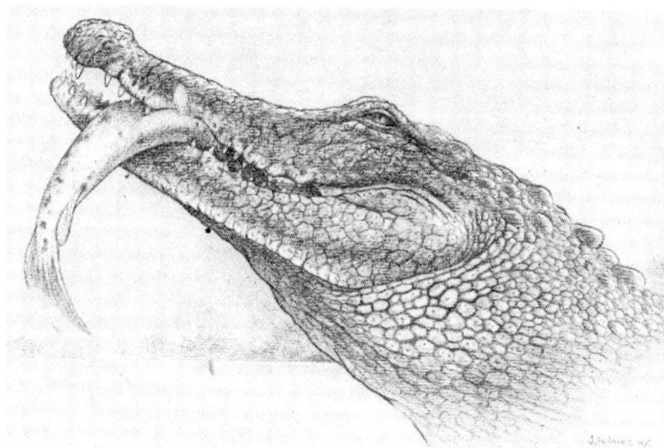

b

Fish scales found inside *Baryonyx'* ribs belonging to this kind of fish called *Lepidotes.*

them in its long jaws. *Lepidotes* grew to about a metre in length so one fish would have made a substantial meal even for a large dinosaur.

Perhaps *Baryonyx* was a scavenger too, the long snout reaching soft tissues inside the body cavity of carrion. It would have been able to breathe while doing so – or while fishing with the end of its snout in water – because the nostril opening was far back from the tip of the snout.

However *Baryonyx* fed, its diet was clearly different from that of other large flesh-eaters like *Allosaurus*. *Baryonyx'* long, low, narrow skull is quite unlike that of *Allosaurus* which has a much broader, higher skull, a wide gape and slicing teeth for dealing with large chunks of flesh.

The heads of *Allosaurus* (a) and *Baryonyx* (b) compared.

Southern England 124 million years ago

The fossilized remains of *Baryonyx* were found in rocks about 124 million years old. The rocks themselves, together with the animal and plant fossils they contain, give clues as to what southern England was like in that distant age. There were steep-sided mountains where London is today. The foothills were forested with conifers, monkey puzzle trees, cycads and ferns. Rivers flowed south to a vast low-lying flood plain that extended as far as France and Belgium.

Baryonyx lived in what is now Surrey but was then near the edge of the flood plain. One hundred and twenty-four million years ago that area was covered by large lakes, rivers and marshes. Plants such as horsetails and ferns grew in abundance, and insects – especially dragonflies – were common. Turtles and crocodiles lived in the water alongside the fishes. The remains of other dinosaurs are found too, the plant-eating *Iguanodon* is the commonest dinosaur from rocks of this age in England and in Belgium. It is known from many skeletons, some of which have been found in the same quarry as *Baryonyx.* Other dinosaurs known from the Wealden include *Megalosaurus, Hypsilophodon, Polacanthus* and sauropods.

A scene in southern England 124 million years ago.

These plant-eating dinosaurs lived in the same area and at the same time as *Baryonyx*.

(a) *Iguanodon,* which grew to 10 metres in length. It had a large thumb spike which it may have used for defence.

(b) *Hypsilophodon* was about 2 metres long. It was agile and fast moving, possibly using its long stiff tail as a stabiliser when running.

(c) *Polacanthus,* an armoured dinosaur equipped with defensive spines along the neck, back and tail and a bony shield over the hips.

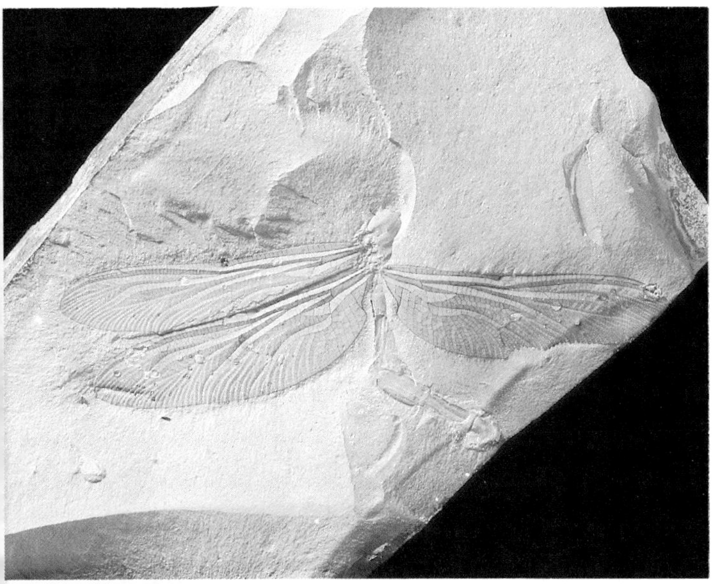

An exceptionally well-preserved hawker dragonfly showing wing veins and body stripes.

Were there any similar dinosaurs?

Deinonychus from the Early Cretaceous of the USA.

Baryonyx is the only reasonably complete skeleton of a large flesh-eating dinosaur found in rocks of Early Cretaceous age anywhere in the world. One small Early Cretaceous theropod, *Deinonychus* from the USA, is known from good skeletons. *Deinonychus* grew up to 2 metres long, with a vicious sickle-like claw on each hind-foot for slashing at prey. The claw-bone of *Deinonychus* is compressed and blade-like, not at all like the large robust claw-bone of *Baryonyx* which was probably borne on the hand.

The remains of other large flesh-eaters from this period are very fragmentary. However, the discovery of *Baryonyx* has enabled us to recognize similar bones collected by French palaeontologists, some 20 years ago, from slightly younger rocks in the Sahara Desert in Niger. Isolated snout fragments and claw-bones, the only pieces that were found, are almost identical to those of the Surrey dinosaur, so *Baryonyx* may have lived over a wide area extending from southern England to the northern half of Africa.

Baryonyx and the Sahara dinosaur... the snouts (a), (b) in side view and from beneath showing the tooth sockets and the claw-bones (c), (d).

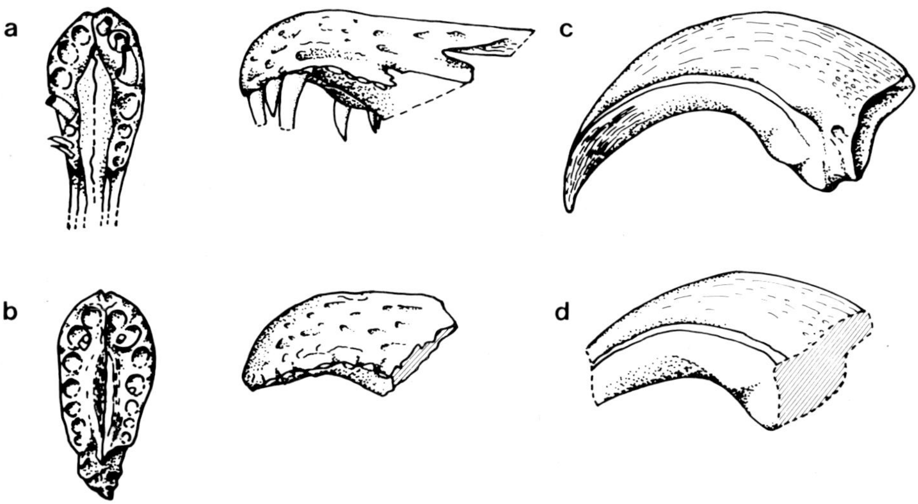